Andrey Amand

Comprehensive functional and physiological rehabilitation

Andrey Amand

Comprehensive functional and physiological rehabilitation

Methods and devices for forming processes of complex functional and physiological rehabilitation

ScienciaScripts

Imprint
Any brand names and product names mentioned in this book are subject to trademark, brand or patent protection and are trademarks or registered trademarks of their respective holders. The use of brand names, product names, common names, trade names, product descriptions etc. even without a particular marking in this work is in no way to be construed to mean that such names may be regarded as unrestricted in respect of trademark and brand protection legislation and could thus be used by anyone.

Cover image: www.ingimage.com

This book is a translation from the original published under ISBN 978-620-7-64054-6.

Publisher:
Sciencia Scripts
is a trademark of
Dodo Books Indian Ocean Ltd. and OmniScriptum S.R.L publishing group

120 High Road, East Finchley, London, N2 9ED, United Kingdom
Str. Armeneasca 28/1, office 1, Chisinau MD-2012, Republic of Moldova, Europe
Printed at: see last page
ISBN: 978-620-7-93619-9

Andrei Amand

Comprehensive functional and physiological rehabilitation

Methods and devices for forming processes of complex functional and physiological rehabilitation using elements of artificial intelligence and artificial neural networks.

Keywords:

Complex rehabilitation; Functional rehabilitation; Physiological rehabilitation; Stress loads; Stress loads of all kinds; Adequate response; Smart production technologies; Mobile communication tools; Non-contact resonance sensors; Real-time monitoring;

Annotation:

In modern society, which implements and organises various variants of complex multifaceted development of innovation economy, especially in areas related to smart industries and technologies, stresses of all kinds arising in the most active organisers of project development processes and generators of new technical and commercial ideas aimed at optimisation and acceleration of development processes, require adequate response and unobtrusive, but extremely reliable and natural technologies and special equipment.

At present, a principle model of system controlling and regulating complex model subsystems has been created using contactless resonant sensors operating on the principles of electromagnetic resonance spectroscopy and having different constructive design as dynamic links between the mobile controlled object and mobile communication means.

Also the principal model of system controlling and regulating complex model subsystems with the use as dynamic links between the mobile controlled object and mobile communication means, made in the form of smart watches, contactless resonant sensors, working on the principles of electromagnetic resonance spectroscopy and having different design.

For so-called smart watches, the sensor design is a - flat coil, - electronic micro - board with the original flat solenoid topology.

Such a sensor receives energy from the watch battery and is constantly in the mode of monitoring the body parameters of a table tennis player.

In this case, real-time monitoring can measure several important parameters that can be affected by the overly intensive nature of the game, e.g. blood sugar concentration, blood pressure, etc.

Content:

Introduction:

Tabletop sports games are one of the most important and promising sources for comprehensive rehabilitation technologies.

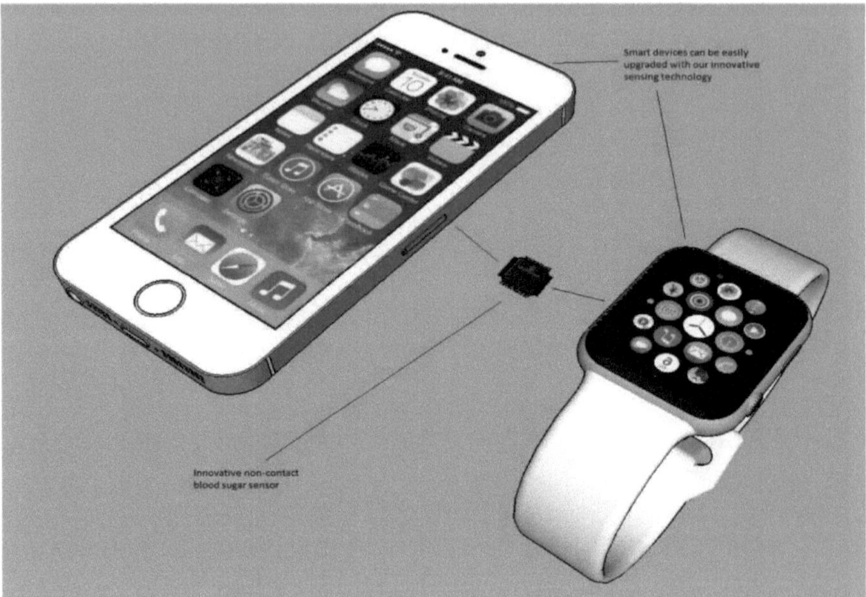

Figure 1, - the figure shows the model of system controlling and regulating complex model subsystems with the use of non-contact resonant sensors operating on the principles of electromagnetic resonance spectroscopy and having different constructive design as dynamic links between the mobile controlled object and means of mobile communication.

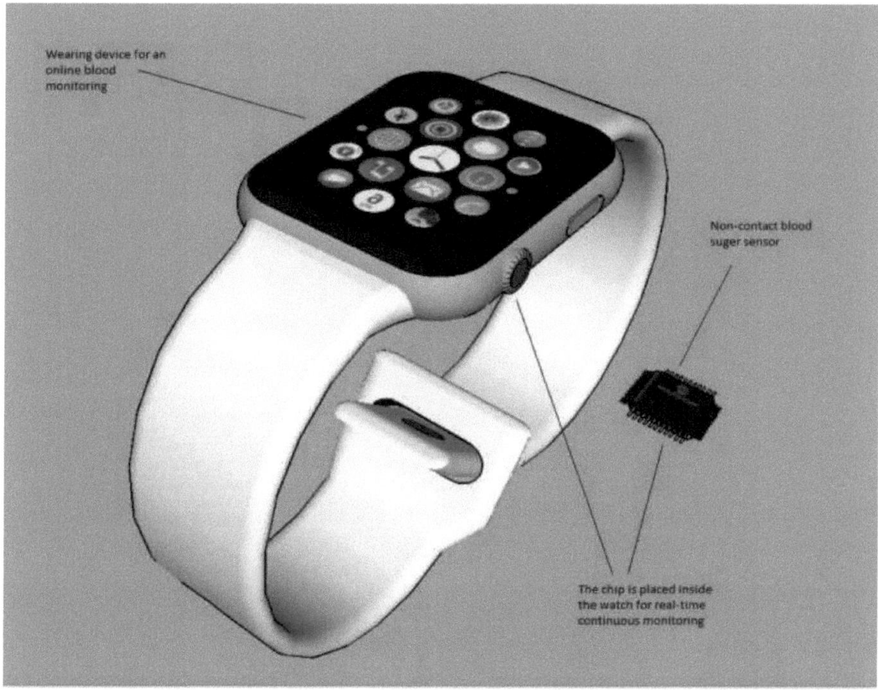

Figure 2,- the figure also shows the model of system control and regulating complex model subsystems using as dynamic links between the mobile controlled object and means of mobile communication made in the form of smart watches, non-contact resonant sensors operating on the principles of electromagnetic resonance spectroscopy and having a different design.

For the so-called smart watches, the sensor design is a flat coil - an electronic micro-board with the original topology of a flat solenoid. Such a sensor receives energy from the watch battery and is constantly in the mode of monitoring the parameters of the table tennis player's body.

In this case, real-time monitoring can measure several important parameters that can be affected by the overly intensive nature of the game, e.g. blood sugar concentration, blood pressure, etc.

These applications become especially important in case of integration of artificial intelligence and artificial neural networks elements into software systems of the whole rehabilitation complex.

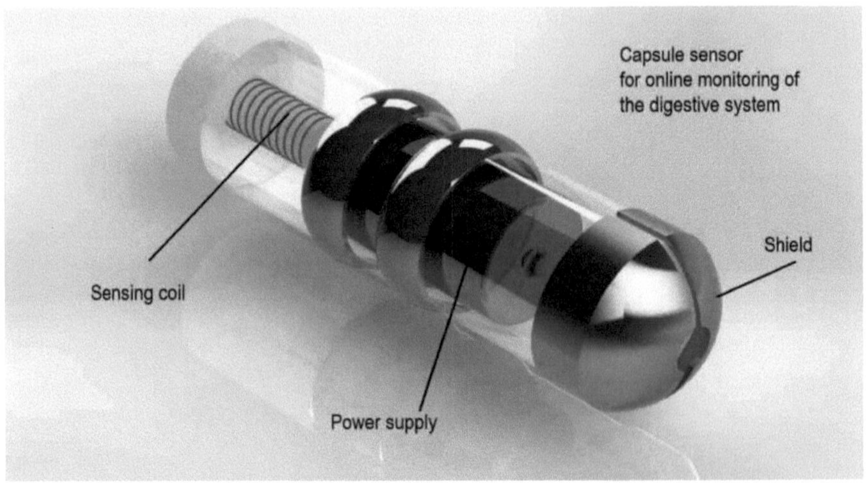

Figure 3, - the figure also shows a model of an integral sensor that is inserted into the axial hole of the handle of a table tennis racket.

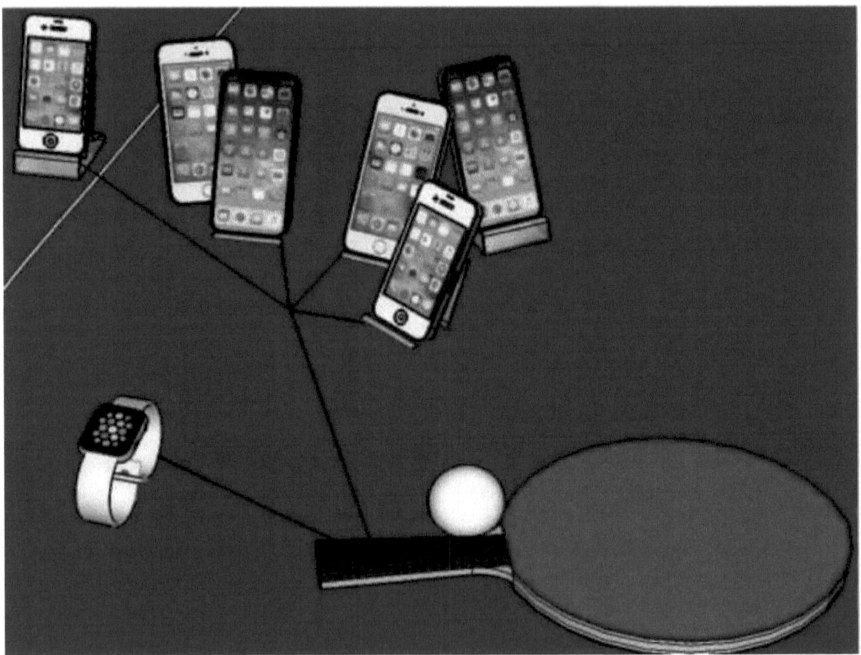

Figure 4, - the figure also shows the interaction model of the integrated sensor with mobile communication objects.

Figure 5, - the figure shows a model of a modular table tennis racket with a software analytical unit linked to a universal mobile autonomous sensor module embedded in an integral handle, whose operation scheme includes elements of artificial intelligence and artificial neural networks.

The invention relates to the field of smart tools and devices and includes a number of interconnected blocks, modules and devices having the ability to collectively or autonomously form a certain structural software system of interaction with table tennis players having each individual connections between themselves and with mobile software devices, mainly in the form of mobile applications, the structure of each of which includes elements of artificial intelligence and artificial neural networks.

In addition, the subject matter of the invention includes, associated with the racket handle, built on the basis of and in development of a structural scheme with realised electromagnetic resonance spectroscopy, stabilising structural elements and basic parts forming cohesive functional structural and technological schemes, built on the basis of conductive properties of carbon-carbon composite fabrics, wherein the carbon-carbon composition is applied by means of high-temperature pyrolysis, and, wherein said fabrics having viscous properties of carbon-carbon composite fabrics, wherein the carbon-carbon composite fabrics are applied by means of high-temperature pyrolysis.

Figure 6, 7, 8, - the figure also shows the modular table tennis racket model with software analysis unit.

A modular racket for table tennis, including a functionally connected impact plate and handle with a cylindrical inner hole and with an integral sensor module installed in this hole, the sensors of which mainly work on the basis of the principles of electromagnetic spectroscopy, and the signals from the sensors in real time mode are fed to analytical software tools with elements of artificial intelligence and artificial neural networks, whereby the impact plate of the racket on both working sides is covered with the following elements

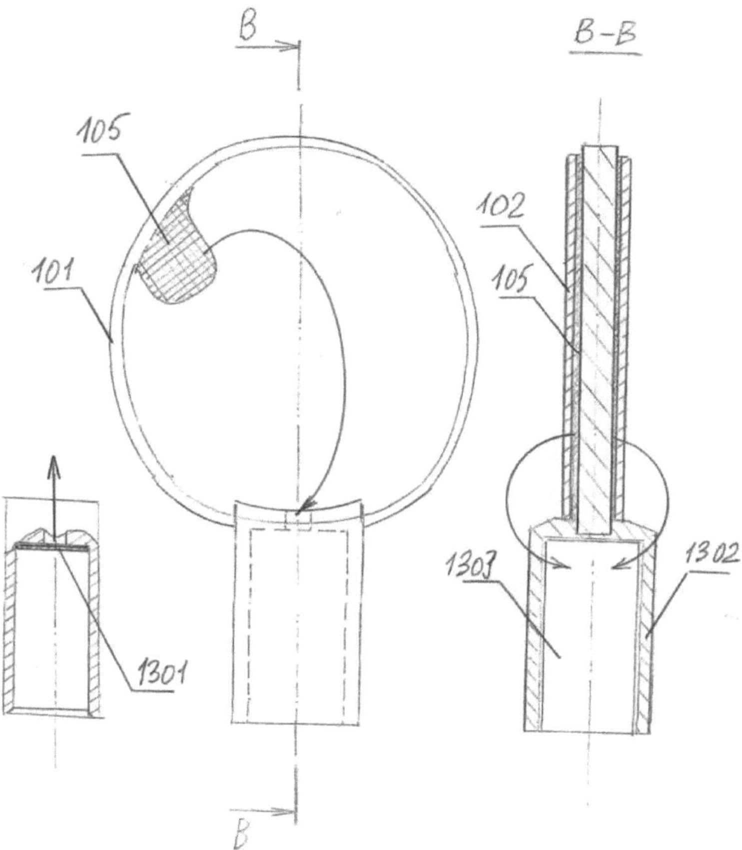

FIG. 1 shows the invention, which relates to devices and methods in the field of intelligent tools and devices and includes a number of interconnected units, modules and devices having the ability to collectively or independently form a certain structural software system of interaction with table tennis players, each of which have individual connections with each other and with mobile software devices, mainly in the form of mobile applications, the structure of each of which includes elements of the art of the invention. In addition, the subject matter of the invention includes, associated with integrated in the racket handle, constructed on the basis of and in development of a structural scheme with implemented electromagnetic resonance spectroscopy, stabilising structural elements and basic parts forming cohesive functional structural and technological schemes based on the conductive properties of carbon-carbon composite fabrics, wherein the carbon-carbon composition is applied by high-temperature pyrolysis and wherein these fabrics, having a viscose woven string and a viscose woven fabric, are applied by high-temperature pyrolysis. The subject matter of the invention includes racket racket racks built on the basis of and in development of a structural scheme with realised electromagnetic resonance spectroscopy, stabilising structural elements and basic parts forming cohesive functional structural and technological schemes built on the basis of conductive properties of carbon-carbon composite fabrics, in which carbon-carbon composition is applied by means of high-temperature pyrolysis and, in this case, these fabrics having viscose woven fabric strings

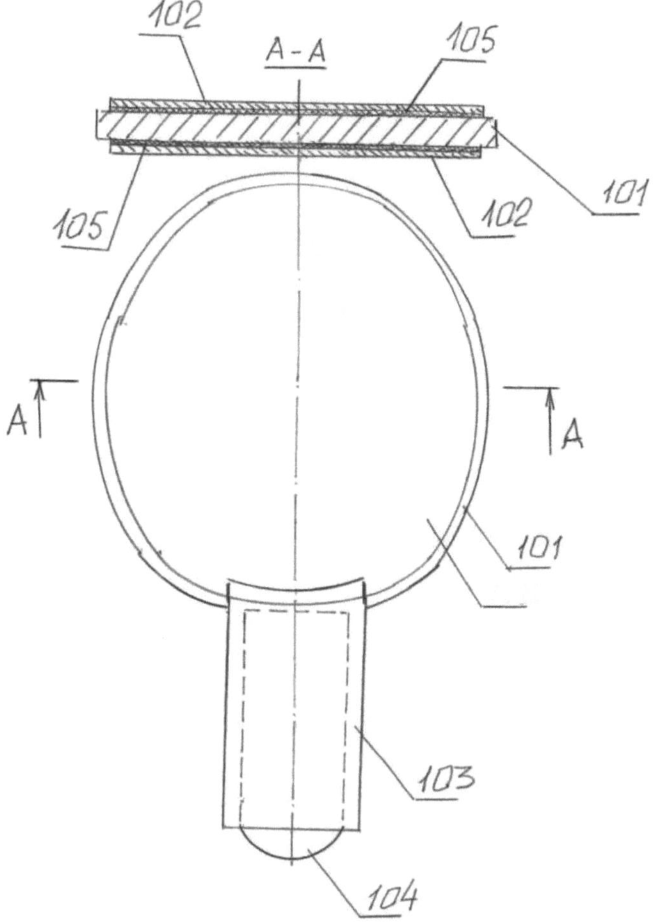

FIG. 2 shows an example of a complex table tennis racket with a combined handle with a cylindrical cavity, in which either a sensor module or analogue thereof, or a weight that turns the racket into a dumbbell is incorporated. The following reference numbers indicate the following features: 101 - Support plate of the table tennis racket, most often made of plywood; 102 - Outer surface of the support plate of the racket, made of rubber or plastic with similar properties; 103 - Body of the racket handle with a hole for the installation of additional modules 104-.

Additional module with ball head; 105 - Carbon cushion - carbon fibre based on viscose fabric.

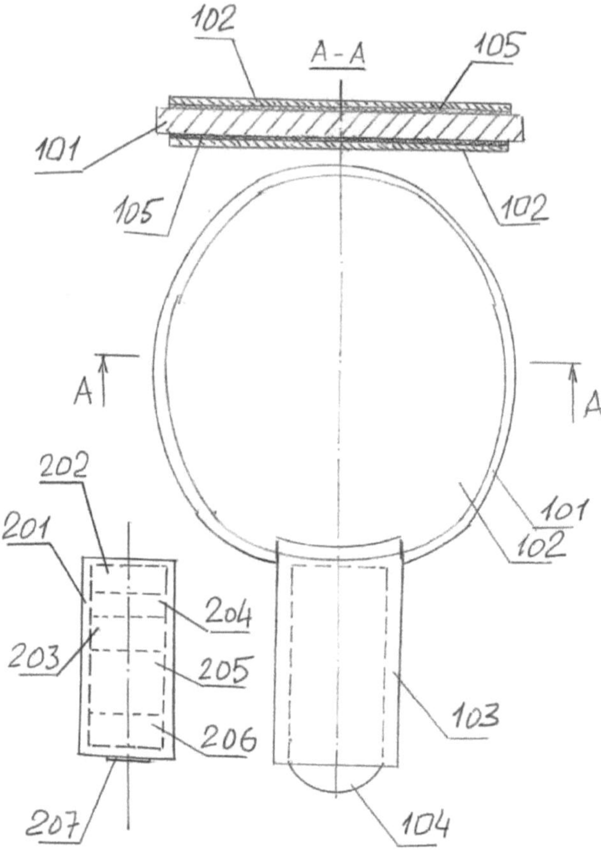

FIG. 3 shows an example of a complex table tennis racket with a combined handle with a cylindrical cavity, in which is embedded either a sensor module or analogue thereof, or a weight that transforms the racket into dumbbells; the figure shows the sensor module

module and its major components. The following reference numbers indicate the following features: 201- Optional table tennis racket handle insert; 202- Electric battery; 203- Primary receiver analyser and resonance signal integrator; 204- Signal analyser and converter; 205- Primary receiver analyser and resonance signal integrator; 206- Pulse generator; 207- Flat coil sensor.

The subject of the invention is a unique racket integrated with sensors in the handle that function based on electromagnetic resonance spectroscopy. The stabilising structural elements and base parts of the racket are made using carbon-carbon composite fabrics treated by high temperature pyrolysis. These innovative materials not only increase the durability and performance of the racquet, but also guarantee a permanent connection to the integrated sensor module. This module, in turn, is able to interact with mobile applications and, through them, with software platforms aimed at sports and rehabilitation. The main objective of this invention is to provide athletes and people in rehabilitation with an effective tool that not only helps to improve physical performance, but also makes the training process more interactive and efficient.

Carbon-carbon composite fabrics are used to create stabilising structural elements and base parts of the racket. These materials are processed by high-temperature pyrolysis, which gives them high strength, lightness and resistance to various external influences. Carbon-carbon composites have high mechanical properties and heat resistance, making them an ideal choice for use in sports equipment where a combination of lightness, durability and reliability is required.

FIG. 4 shows an example of a complex table tennis racket having a combined handle with a cylindrical cavity, in which either a sensor module or an analogue thereof is incorporated, or a weight that transforms the racket into a dumbbell. The figure shows a racket with a handle in which a weight is integrated, turning the racket into a dual-function instrument or a multi-disciplinary sports equipment. The following reference numbers denote the following features: 301- Ball-headed racket handle insert; 302- Special friction spring rings on the outside diameter of the table tennis racket handle; 303- Special friction spring rings on the outside diameter of the table tennis racket handle; 304- Lower flange for the table tennis racket handle; 305- Cylindrical insert in the table tennis racket handle; 306- Special friction spring rings on the outside diameter of the table tennis racket handle; 307 - Contact end of the insert in the handle of the table tennis racket; 308 - Special friction spring rings on the outside diameter of the handle of the table tennis racket; 309 - Special friction spring rings on the outside diameter of the handle of the table tennis racket; 310 - Special friction spring rings on the outside diameter of the handle of the table tennis racket.

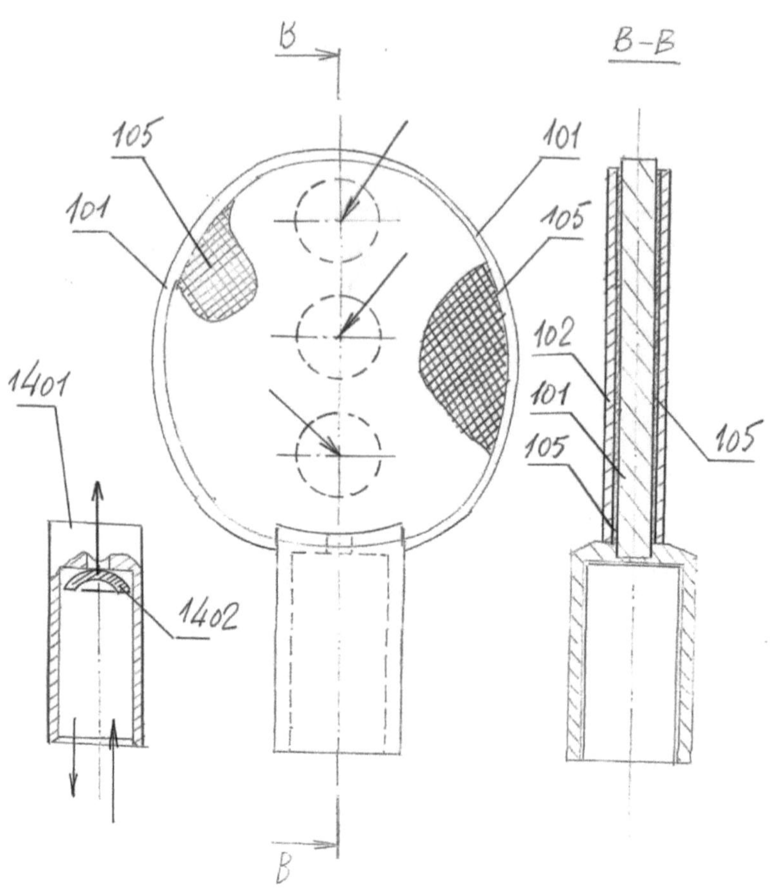
B
B–B
105
101
101
105
102
101
105
105
1401
1402
B

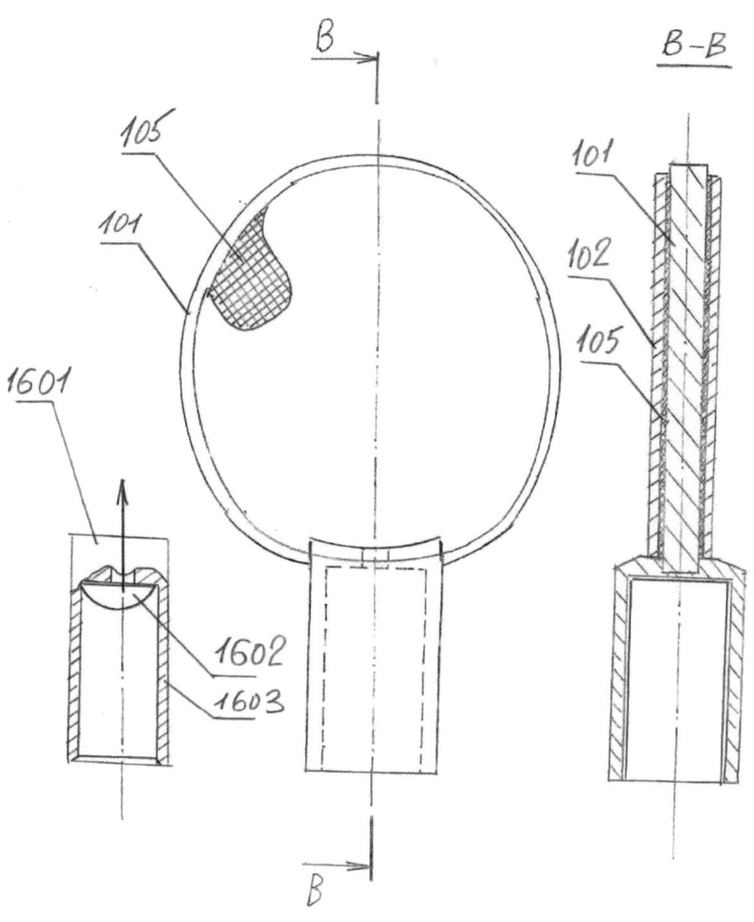
B
B-B
105
101
101
102
105
1601
1602
1603
B

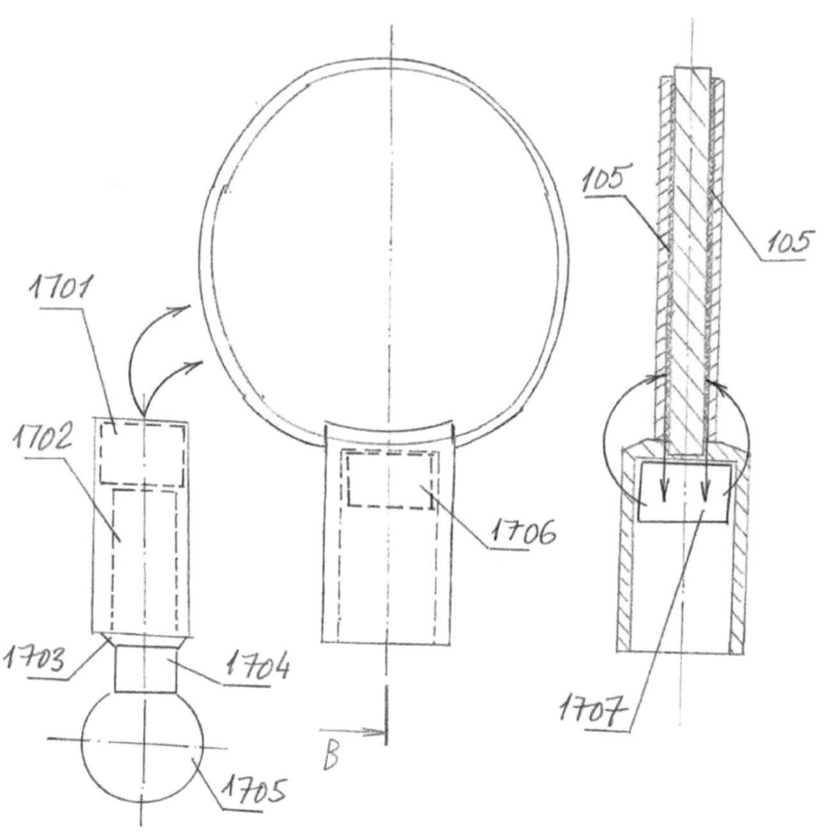
1701
1702
1703
1704
1705
1706
B
105
105
1707

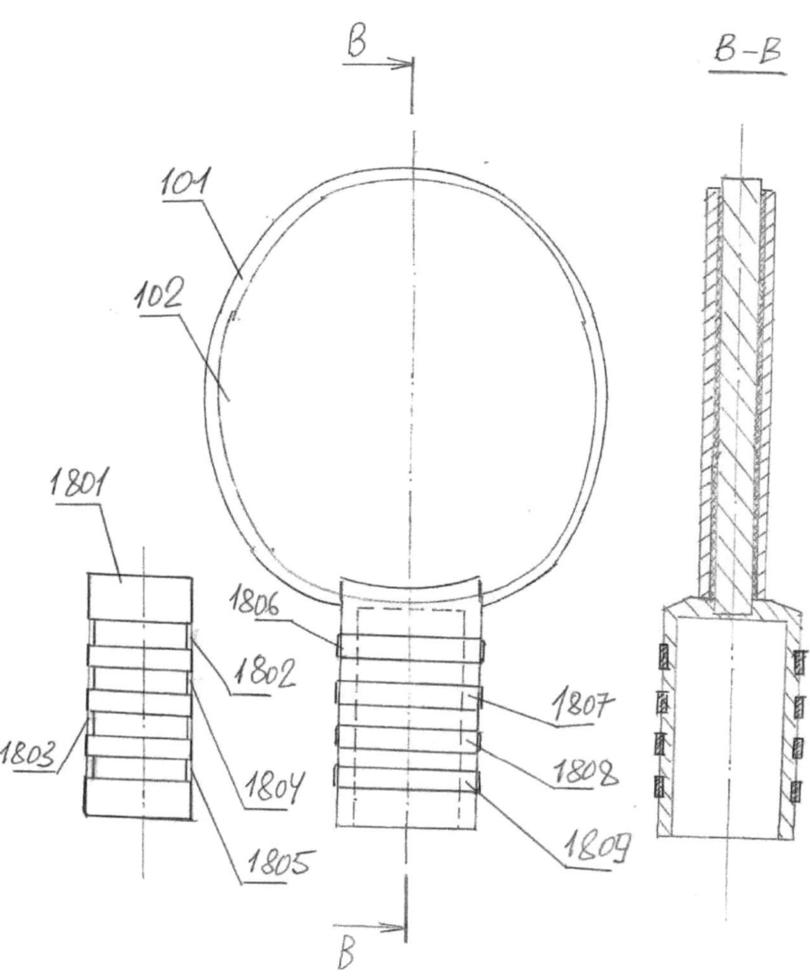
B
B-B
101
102
1801
1806
1802
1807
1803
1808
1804
1809
1805
B

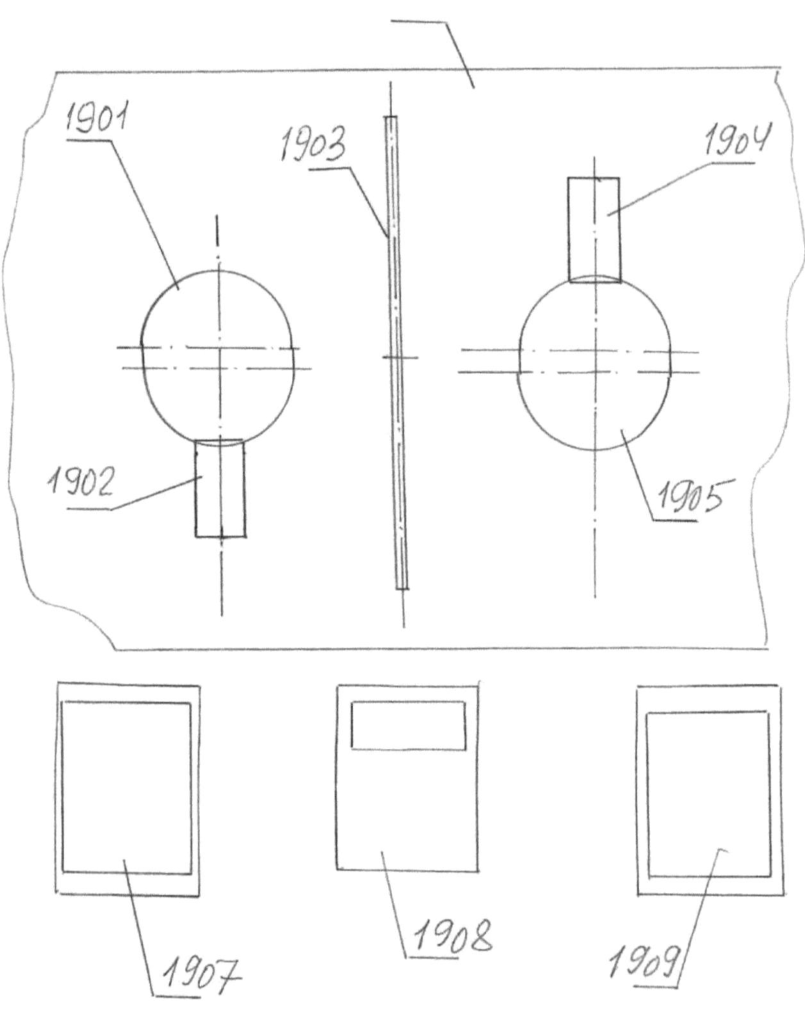
1901
1903
1904
1902
1905
1907
1908
1909

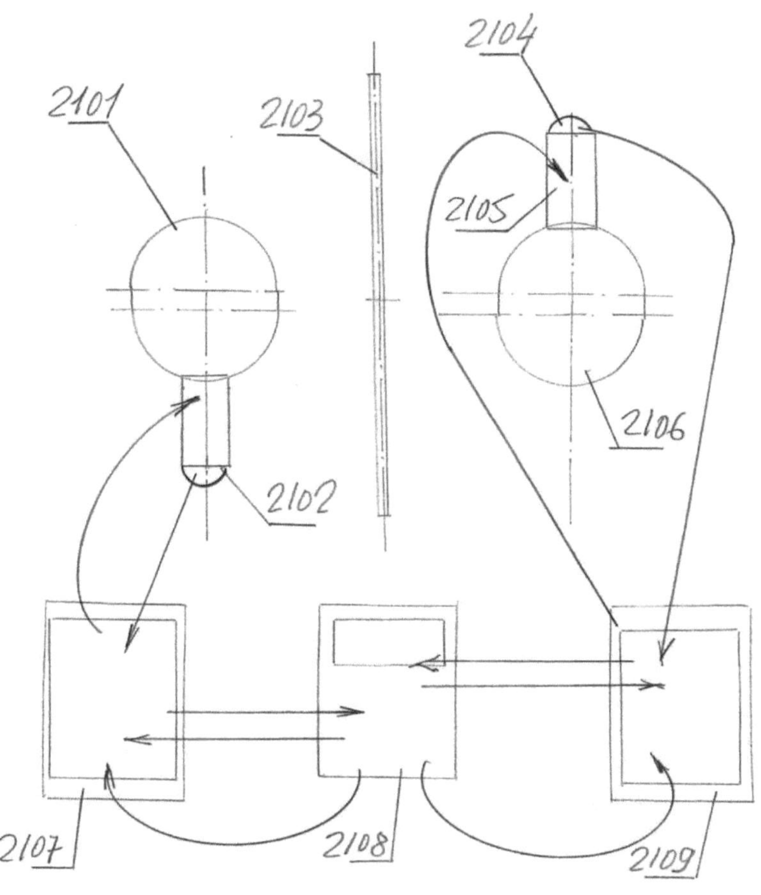
2101
2103
2104
2105
2106
2102
2107
2108
2109

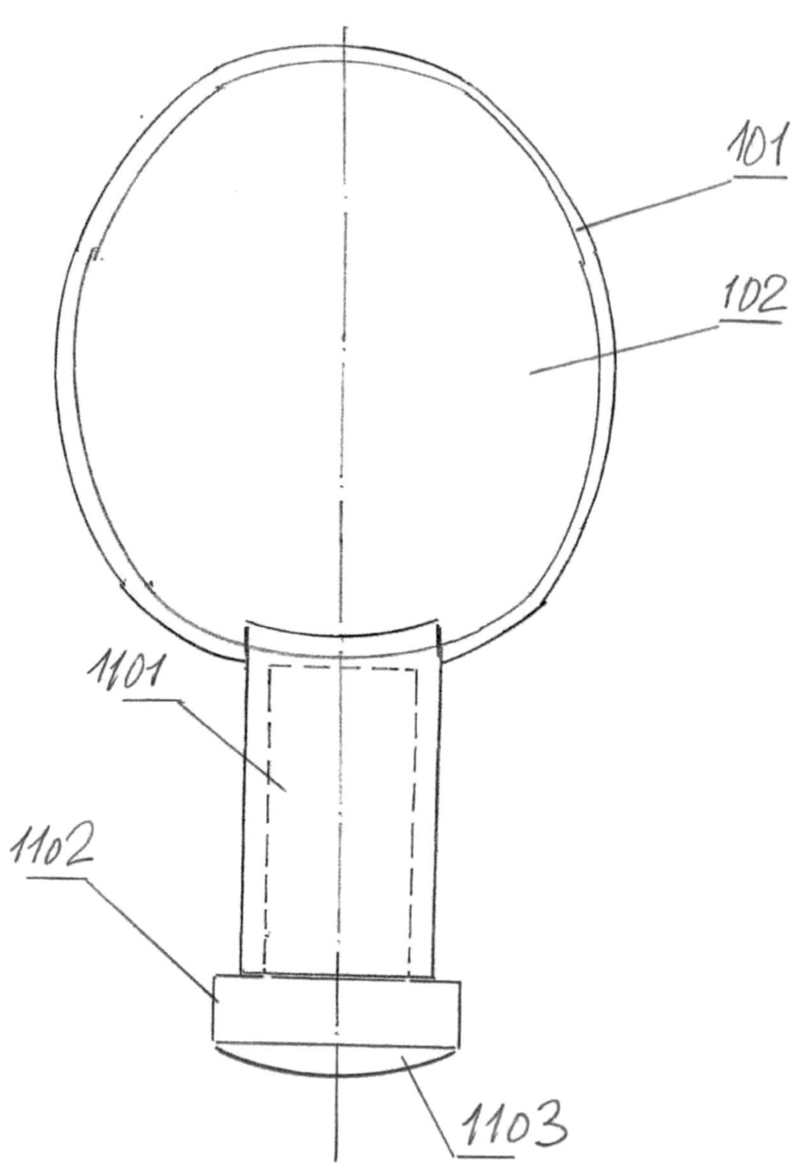
101
102
1101
1102
1103

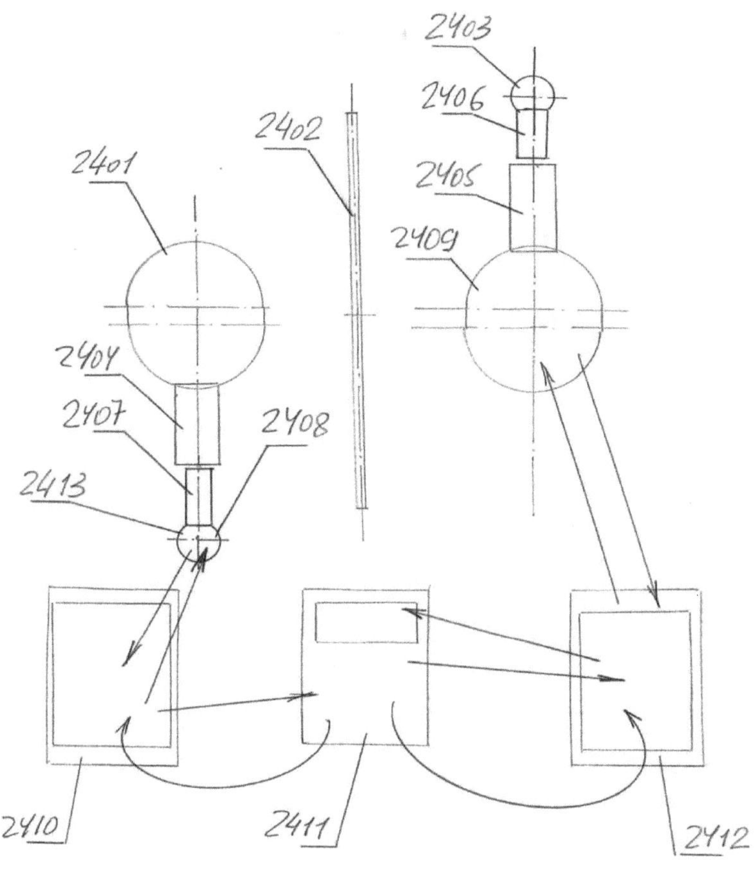
2403
2406
2402
2401
2405
2409
2404
2407
2408
2413
2410
2411
2412

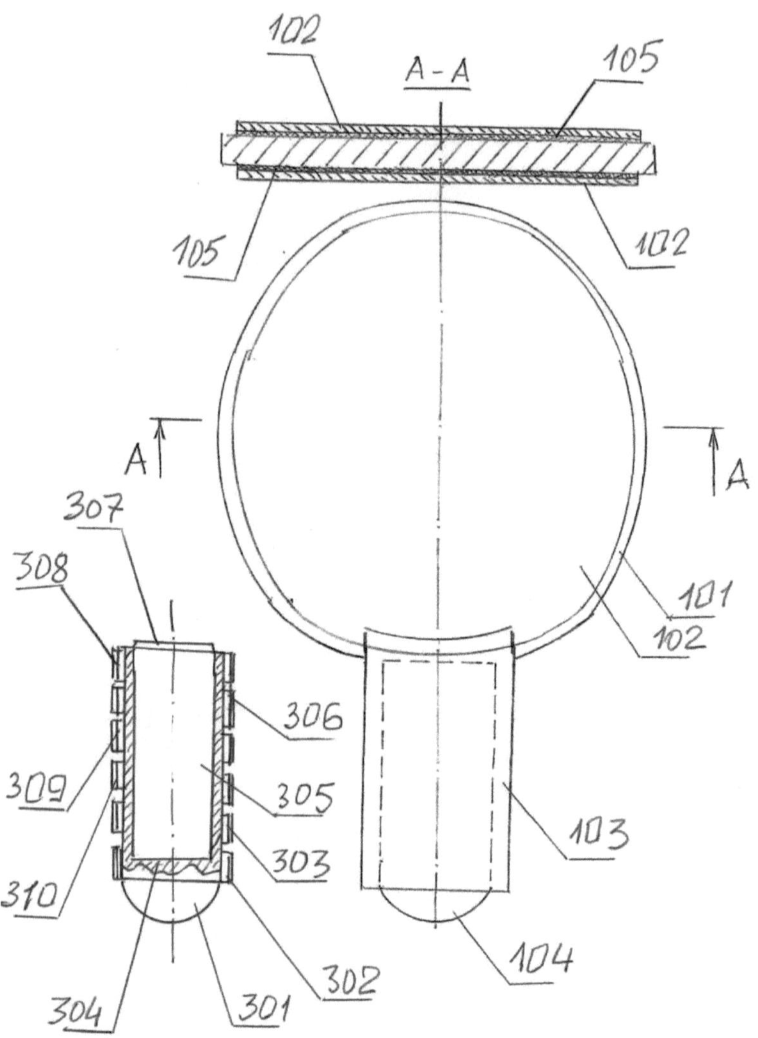
102
A - A
105
105
102
A
A
307
308
101
102
306
309
305
103
303
310
302
104
304
301

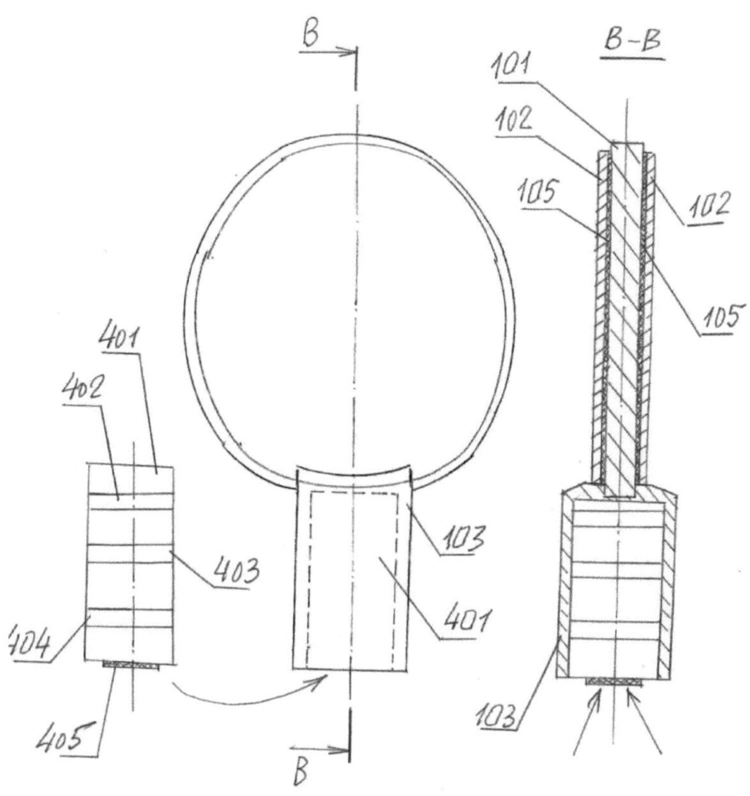
B
B-B
101
102
102
105
105
401
402
403
103
401
404
405
103
B

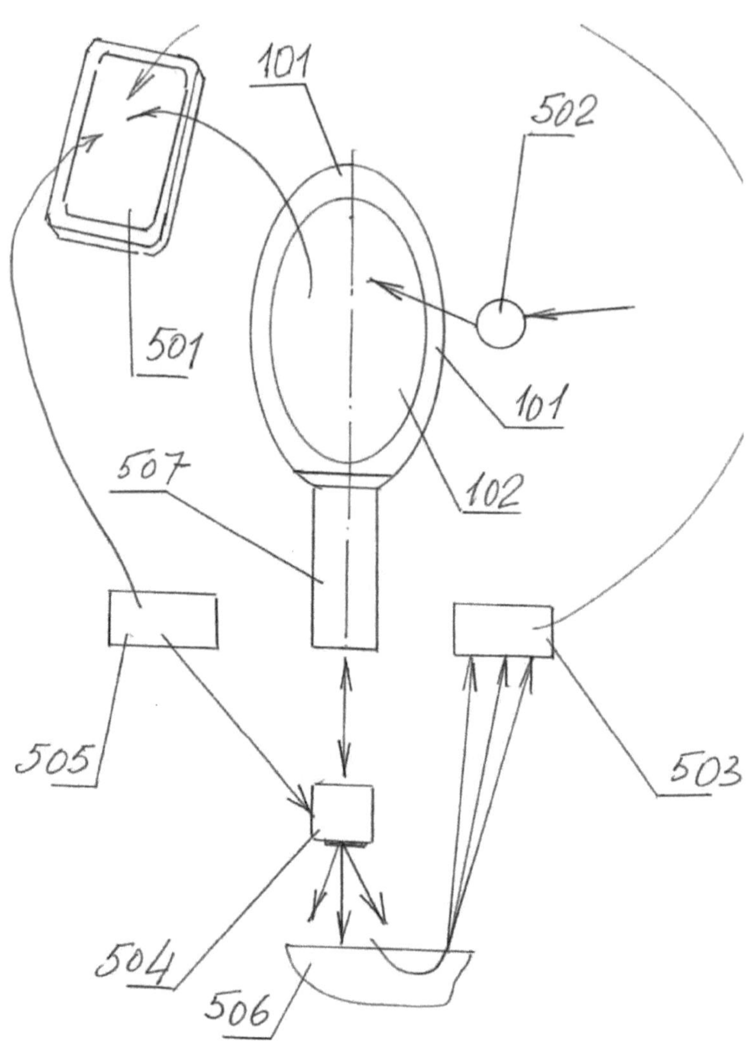
101
502
501
101
507
102
505
503
504
506

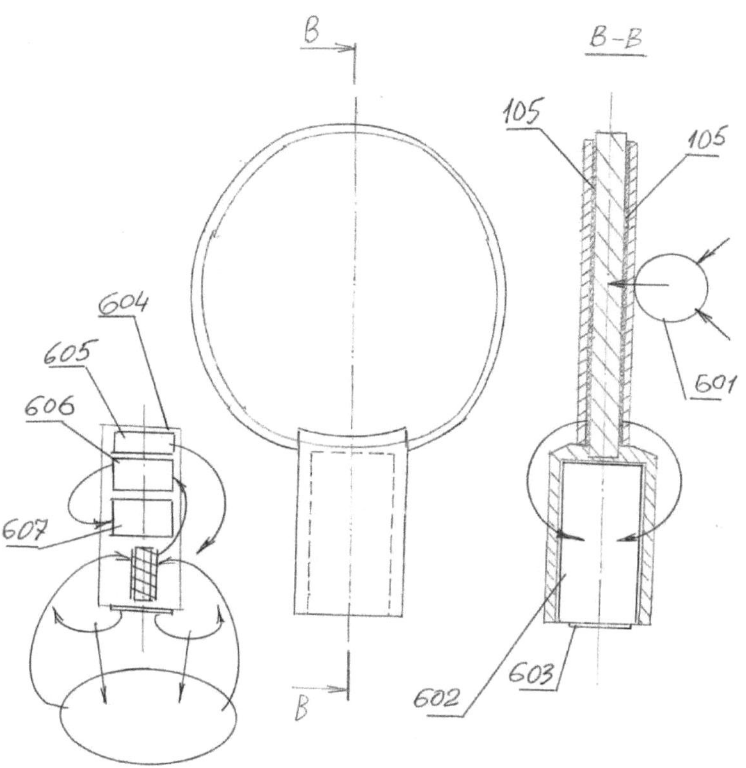
B
B-B
105
105
601
604
605
606
607
B
602
603

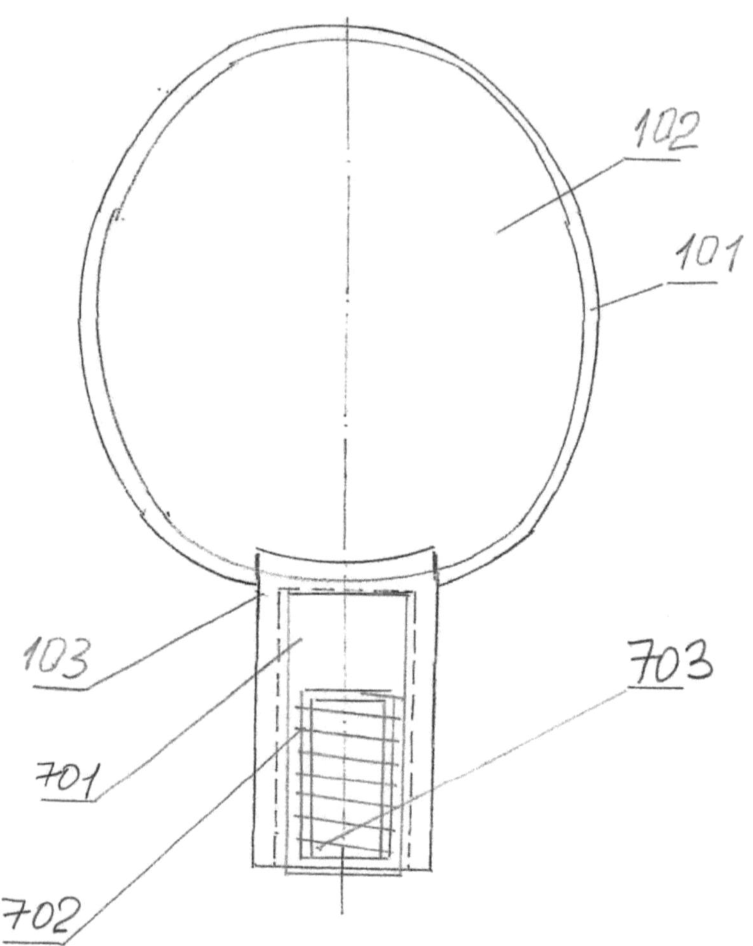
102
101
103
703
701
702

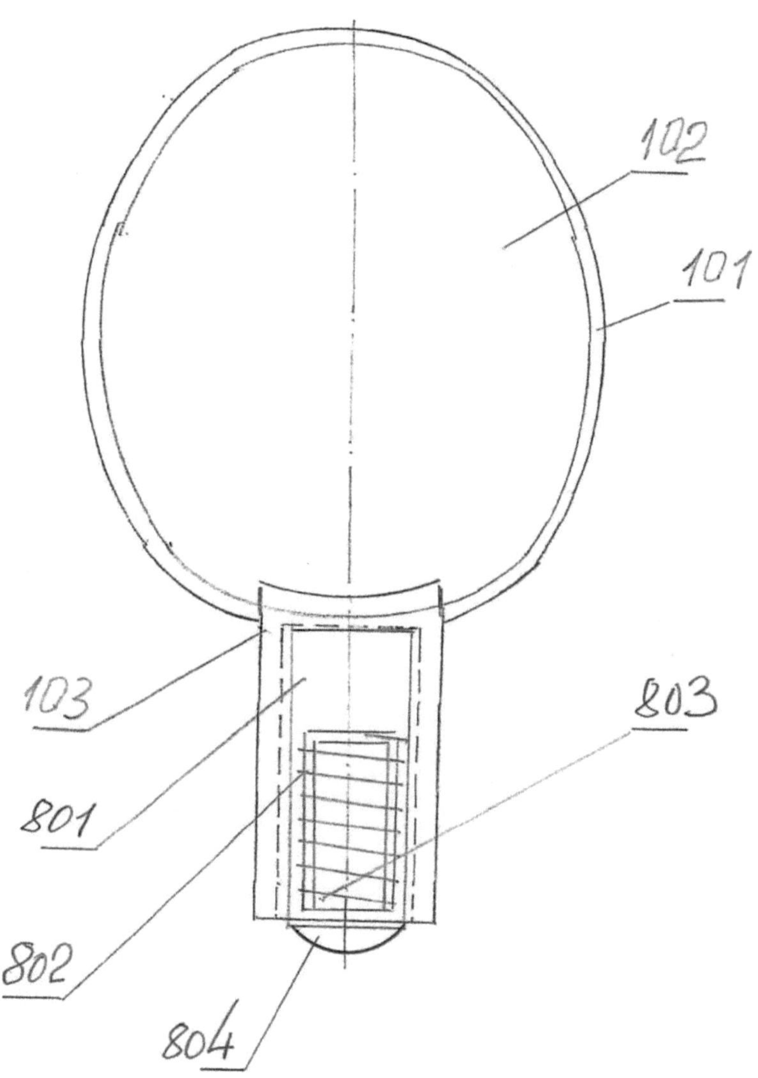
102
101
103
803
801
802
804

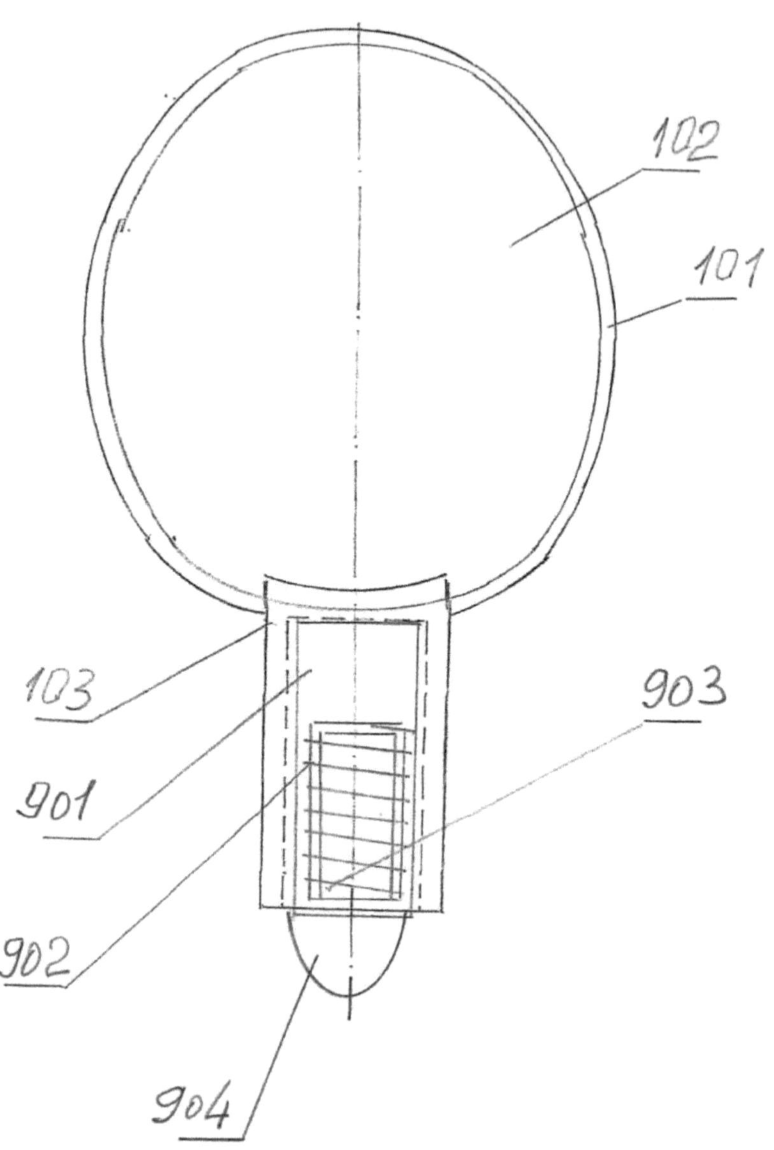
102
101
103
903
901
902
904

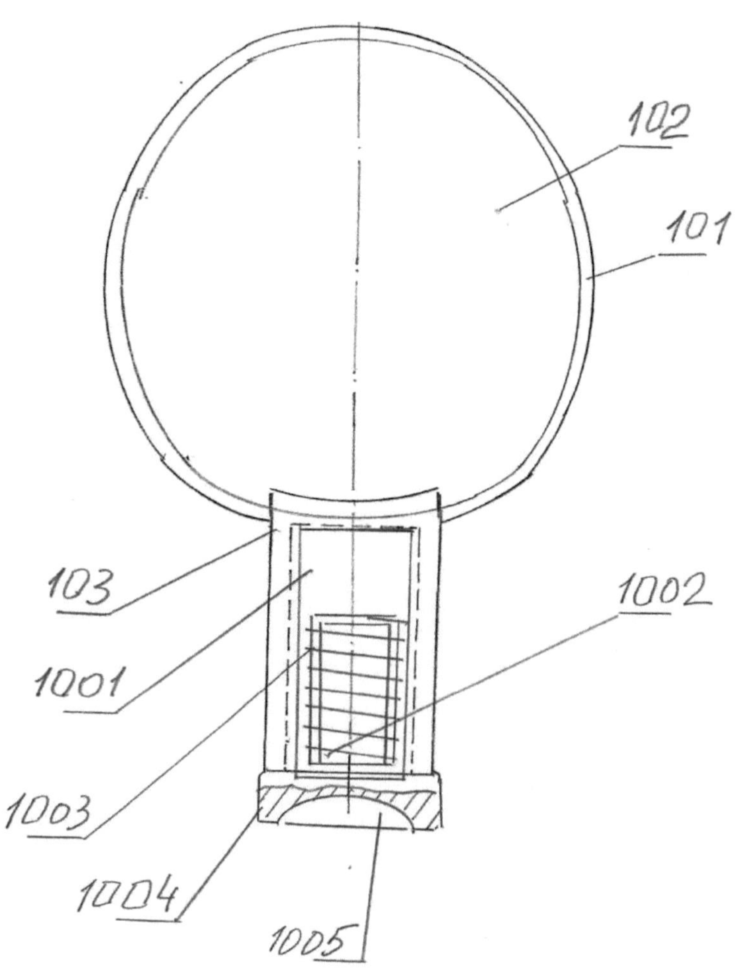
102
101
103
1002
1001
1003
1004
1005

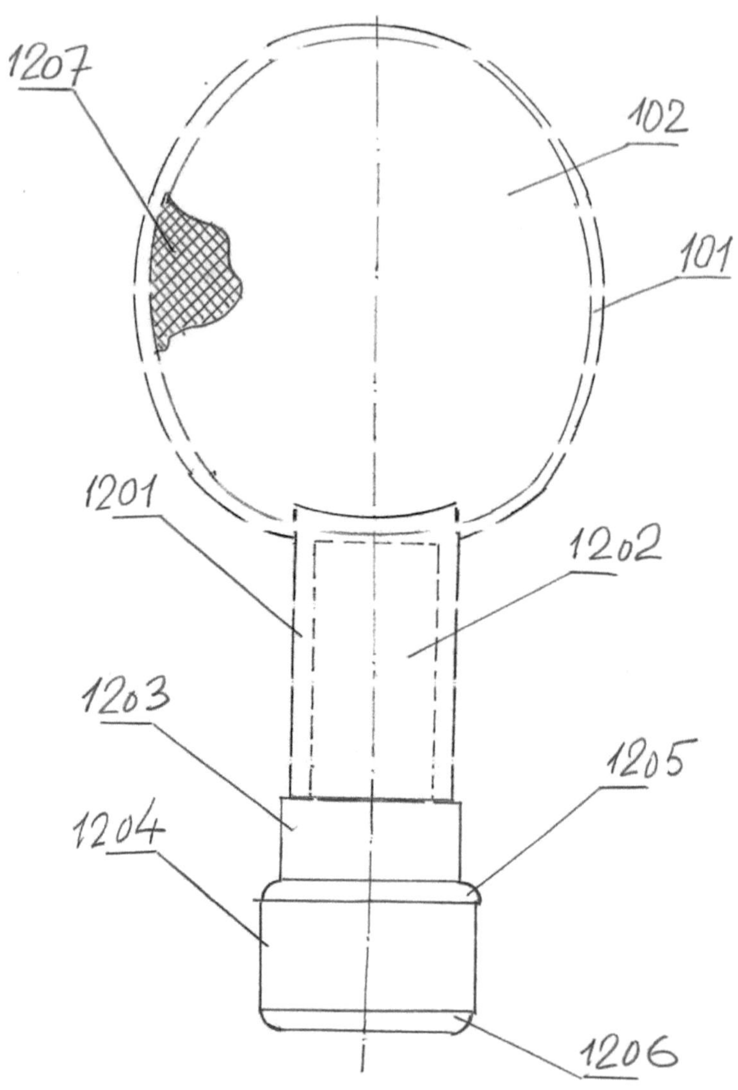
1207
102
101
1201
1202
1203
1205
1204
1206

LIST 01 OF LITERATURE, PATENT AND LICENCE INFORMATION USED

APPENDIX 1

United States Patent10 ,065,068

WilsonSeptember 4, 2018

Adjustable ankle rehabilitation apparatus

Abstract

Various embodiments provide an adjustable ankle *rehabilitation device* for rehabilitating torn ligaments associated with a sprained ankle. The *rehabilitation device* can include a planar platform secured to a shoe, and a balancing rail adjustably attached to the bottom of the platform and extending fore to aft. The balancing rail is configured to selectively place a desired amount of *stress* on the medial muscle or, alternatively, the lateral muscle by adjusting the balancing rail from side-to-side. The *device* can include adjustable fasteners to secure the balancing rail at a desired position adjacent the bottom of the platform.

APPENDIX 2

United States Patent 9,616,283

Heinicke, et al. April 11, 2017

Therapeutic device

Abstract

A low *stress* therapeutic *device* is provided by utilizing foot plates and guide rails having operationally tracking surfaces of a low coefficient of friction supported by a platform. The *device* includes a rail stabilizer equipped with extending longitudinally recess or slot and a slidably mounted foot plate having upon its underside a longitudinal projection slidably retained within the rail recess. The therapeutic *device* may be designed to operate under relatively effortless strain at a low coefficient of friction. The therapeutic *device* is useful for knee replacement, stroke victims, ACL repair, and other therapeutic treatments requiring a nominal initiating effort of movement for *rehabilitation.* The *device* may be provided as a one or two footed *device of* a light weight particularly useful in a patient sitting or lying position. The foot plates may be appropriately equipped with longitudinally underside tracking guides reciprocating sliding within longitudinal slots provided by a tracking rail.

ANNEX 3

United States Patent 9,532,916

Tsoi, et al. January 3, 2017

Wearable power assistive device for hand rehabilitation

Abstract

A wearable power assistive *device* for hand *rehabilitation* includes a hand brace having an external platform and an internal platform connected to and spaced inwardly from the external platform. Five finger assemblies are adjustably mounted on and extending from the distal end of the external platform. Each finger assembly includes a proximal follower assembly for a metacarpophalangeal joint. Five motors are used to actuate the five finger assemblies respectively. Each motor is mounted in close proximity to the external platform and has one end connected to the external platform and another end coupled to its proximal follower assembly by a ball joint in order to facilitate transfer of force and minimize mechanical *stress* on the other parts of the *device.*

APPENDIX 4

United States Patent7 ,255,619

RasmussenAugust 14, 2007

Variable resistance aquatic device and methods of using the same

Abstract

An aquatic *device* is usable in an aquatic environment for a variety of purposes, such as physical therapy, *rehabilitation,* and/or exercise. The aquatic *device* permits a person to simulate a walking or running gait cycle in the aquatic environment, reducing the *stress/strain* associated with walking or running on the ground. An aquatic *device* includes a foot-receiving member rotationally coupled to a fin member. The fin member, when in an extended position, provides increased resistance as the person attempts to walk or run in the aquatic environment. During a walking or running gait, the fin member moves into a folded position, thus reducing the resistance of the water on the aquatic *device.* The aquatic *device* is adaptable and modifiable to have varying shapes, designs, sizes, resistance levels, and/or other aspects.

APPENDIX 5

United States Patent6 ,056,613

PikeMay 2, 2000

Multi-purpose floatation device for recreation, exercise, instruction and rehabilitation purposes

Abstract

A recently popular form of exercise and therapy, aquatic exercising devices present unique operating conditions to the body because of their use of water resistance and their buoyancy. By making proper use of water resistance, such devices can provide the body with excellent muscular and cardiovascular training, at the same time, the buoyancy offered by these devices eliminates the *stress* and injuries associated with the jarring impact of such land-based exercises as running and aerobics. It is also an object of the present invention to provide an aquatic exercise devise that is a singular unit. The inventor began attending a water aerobic class in 1995 for health reasons. Exercising in the water took most of the pain out of the movement, but the inventor found that she was still hurting herself. She sought to reach a truly weightless state in which to condition her body. She tried the various devices provided by the pool facility but none proved effective in granting her the non-impact workout she was determined to find. With a problem to solve the inventor experimented, altered and designed a new and improved floatation *device* that is uniquely different in its adaptability to numerous applications. A uniquely different floatation *device* this invention goes beyond the restrictive designs of prior art designed to address one or another aspect of aquatic safety, exercise, *rehabilitation* or recreation. This invention adapts to use in a multitude of expressions from water yoga, a unique synergy of ancient eastern culture and modern-day technology; to aqua aerobic exercises incorporating cardiovascular enhancement activities; *rehabilitation of* physical injury or illness; as well as addressing the basic aspects of water safety and learning how to swim. A floatation *device* for various exercises, instruction, *rehabilitation,* therapeutic and/or recreational purposes; this invention provides floatation support like no other product on the market because of its unique design and flexibility and the multiple number of ways in which it can be used. With this invention its possible to float supine, moving through various water yoga relaxation movements and stretches; ride it like a bicycle seat; sit on it like a swing; wrap it around the torso and clip it on for deep water workout and/or for those who are uncomfortable in water but who must get in for health and/or *rehabilitation* purposes; hold it with hands; slip it under arms, front to back, or back to front; all to move through various exercises for health, *rehabilitation* and fun. The variation is used to provide superior floatation in a clip-on style. With this invention secured around the torso, up the chest and around the back of the neck, the wearer is provided with nohands support. While wearing the invention the wearer can float forward to swim and learn strokes; tread water in an upright position; and/or float supine; all with complete range of motion of limbs and/or torso. This variation of the invention can be utilized in swimming instruction, pool safety, *rehabilitation*, recreation, instruction and general poolside safety.

ANNEX 6

United States Patent | 5,476,429

Bigelow, et al. | December 19, 1995

Treadmill for use with a wheelchair

Abstract

An exercise *device* for the occupant of a wheelchair acting as a treadmill which may be used for cardiac *stress* testing, cardiac or stroke *rehabilitation*, fitness training, aerobic training or educational/physical games, with the *device* including a generally inclined ramp having parallel sides, a forward entrance portion, a movable dolly mounted on rails on the sides of the ramp, the dolly having a pair of laterally movable caster capture plates with openings to receive the front casters of a wheelchair and angular rods cooperating with the wheelchair drive wheels acting to adjust the lateral spacing of said plates, locking means for the dolly to retain it in its forward position, separate locking means for locking the dolly in its rearward position when a wheelchair has been moved onto the ramp into operative position, a pair of enlarged openings adjacent the rear edge of the ramp, and a pair of longitudinally movable rollers beneath the ramp and movable between a rear retracted position allowing the wheelchair drive wheels to be partially received in the openings and a forward position under the drive wheels to engage and lift the drive wheels so that the user can manually rotate the wheelchair drive wheels to rotate the rollers and provide signals to a control apparatus for the desired type of training, testing or *rehabilitation*.

ANNEX 7

United States Patent Application	20130261514
Kind Code	A1
TSUI; Michael Kam Fai; et al.	October 3, 2013

WEARABLE POWER ASSISTIVE DEVICE FOR HAND REHABILITATION

Abstract

A wearable power assistive *device* for hand *rehabilitation* includes a hand brace having an external platform and an internal platform connected to and spaced inwardly from the external platform. Five finger assemblies are adjustably mounted on and extending from the distal end of the external platform. Each finger assembly includes a proximal follower assembly for a metacarpophalangeal joint. Five motors are used to actuate the five finger assemblies respectively. Each motor is mounted in close proximity to the external platform and has one end connected to the external platform and another end coupled to its proximal follower assembly by a ball joint in order to facilitate transfer of force and minimize mechanical *stress* on the other parts of the *device*.

ANNEX 8

United States Patent Application 20120329611

Kind Code A1

Bouchard; Marc; et al. December 27, 2012

Motorised Lower Body Rehabilitation Device and Method

Abstract

Disclosed is a motorised *rehabilitation* apparatus and method for disabled, impaired or injured individuals, which trains a proper gait, increases blood flow, relieves *stress,* and reconditions lower body muscles and joints. The *device* comprises a powered stationary bicycle having a seat, handle grips, and rotating foot pedals that receive motive input from an electric motor and user input. The *device* further includes a pair of thigh braces that are connected together between the user's thighs via a hinge able link and chain that controls and trains an individual's limbs through the pedal rotation. The disclosed method further combines the present bicycle *device for rehabilitation* in conjunction with visual stimuli in the way of a three-dimensional television display that stimulates endorphins, relieves mental *stress* and allows the motive input from the bicycle and mild user input to exercise the limbs of a user without focusing on the *rehabilitation* activity.

ANNEX 9

United States Patent Application	20070093153
Kind Code	A1
Rasmussen; Scott K.	April 26, 2007

Variable resistance aquatic device and methods of using the same

Abstract

An aquatic *device* is usable in an aquatic environment for a variety of purposes, such as physical therapy, *rehabilitation,* and/or exercise. The aquatic *device* permits a person to simulate a walking or running gait cycle in the aquatic environment, reducing the *stress/strain* associated with walking or running on the ground. An aquatic *device* includes a foot-receiving member rotationally coupled to a fin member. The fin member, when in an extended position, provides increased resistance as the person attempts to walk or run in the aquatic environment. During a walking or running gait, the fin member moves into a folded position, thus reducing the resistance of the water on the aquatic *device.* The aquatic *device* is adaptable and modifiable to have varying shapes, designs, sizes, resistance levels, and/or other aspects.

ANNEX 10

United States Patent Application 20060211937

Kind Code A1

Eldridge; Robert September 21, 2006.

Garment to facilitate use of a portable monitor device

Abstract

A garment configured to hold a portable medical *device,* and more particularly to a modified upper garment to hold, secure and conceal a heart monitor while allowing easy and unobtrusive access to cardiac lead points on a patient. The garment has an exterior pocket for a monitor. It further has a plurality of openings to allow attachment of monitor leads on a patient without the need to remove the garment. The openings may also have closure means. The garment provides modesty, comfort, durability and an attractive appearance. The garment may be configured for use in all cardiac *rehabilitation* situations including exercise and *stress* testing. The entire garment is made of x-ray transparent materials.

ANNEX 11

United States Patent Application	20060142680
Kind Code	A1
Iarocci; Michael Anthony	June 29, 2006

Active assistance for the ankle, knee and other human joints

Abstract

A human joint assist *device* that applies a torque at the joint to assist physiological exertion forces, which is the load carrying task of the joint and surrounding muscles, tendons, and ligaments. The application of this *device* reduces the physiological exertion force requirement, and may be adjusted with respect to assist level, to suit the issue associated with joint motion and is useful for joint *rehabilitation* and sports activities. Among other things, this results in a reduction of physiological exertion force in a fashion that makes it easier to extend the levers (long bones) associated with extension against a given resistance. For example, standing from a squatted position with the assistance of this *device* reduces the *stress* on physiological members associated with joint articulation.

ANNEX 12

United States Patent Application	20180001172
Kind Code	A1
SUTTA; Peters; et al.	January 4, 2018

STRUCTURE OF ACCESSORY ELEMENT FOR EQUIPMENT OF FLOORBALL TRAINING COURT AND USE OF IT FOR FORMATION OF FLOORBALL SIMULATOR

Abstract

The invention refers to the equipment of the training rink for floorball, manufacturing of exerciser structural element, applying concept of *tennis* racket stringing. Proposed design of subsidiary element for floorball rink arrangement characterised with that it is made as latticework formed by: two parallel end plates; several threaded rods as stiffening members; two elastic string structures disposed in two parallel planes, at that each of it presents one side of mentioned latticework and provided with: --holes for fastening threaded rods which assure rigidity and load bearing capacity of the subsidiary element frame structure; --holes for crisscross stringing in two parallel planes and string fastening at mentioned end plates independently one from another.

ANNEX 13

United States Patent Application	20160296815
Kind Code	A1
Pindaric; Michael	October 13, 2016

More To Bouncing Ball

Abstract

Easy to assemble and disassemble gaming apparatus that allows a single player to play a game comparable to *Tennis* and/or Ping Pong in a limited space environment The proposed game apparatus also allows a single player to perfect his or her skill.

ANNEX 14

United States Patent Application	20070238561
Kind Code	A1
Hu; Liang-Fa	October 11, 2007

Structure of toy tennis racket

Abstract

Structure of toy *tennis* racket, which mainly improves composition of hitting face of toy *tennis* racket; it stretches string that one side is adhesive across holes around head of *tennis* racket in horizontal and longitudinal to make network, so that one side of this network is adhesive face and another side is the hitting face; such combination makes hitting face of toy *tennis* racket that can produce rebound force owing to the flexible network, furthermore, such racket may provide best ventilation effect to reduce wind resistance, you can hit ball easily as if playing with real *tennis* racket.

LIST 02 OF USED LITERATURE, PATENT AND LICENCE INFORMATION

United States Patent Application	20130158939
Kind Code	A1
YAMAMOTO; Yosuke	June 20, 2013

METHOD AND COMPUTER-READABLE STORAGE MEDIUM FOR FITTING *TENNIS* RACKET AND ANALYSIS DEVICE

Abstract

A *tennis* racket fitting method according to the present invention comprises: a first step of preparing a plurality of test *tennis rackets,* at least one type of racket property that influences swing for hitting a ball being defined for each of the test *tennis rackets*; a second step of causing a reference *tennis* racket to be swung at least one time by a user in order to hit a *tennis* ball, and acquiring a measured value by measuring transition in at least one of the position, the velocity, the acceleration, and the angular velocity of the reference *tennis* racket in at least a portion of an interval from when the swing starts until when the swing ends; a third step of calculating at least one evaluation indicator to be an evaluation of the swing of the reference *tennis* racket based on the measured value; and a fourth step of, based on the racket property, selecting a *tennis* racket that can improve at least one evaluation indicator from among the test *tennis rackets*.

Inventors: YAMAMOTO; Yosuke; *(Kobe-she, JP)*

Applicant: Name City State Country Type

DUNLOP SPORTS CO. LTD.; Kobe-she JP

Assignee: DUNLOP SPORTS CO. LTD.

Kobe-she JP

Family ID: 47355804

Appl. No.: 13/651825.

Filed: October 15, 2012

United States Patent 4,293,129.

Planks October 6, 1981

Game rackets and paddles with non-parallel playing surfaces

Abstract

This relates to game rackets, such as *tennis* rackets, and to paddles, such as *table tennis* paddles, with two discrete playing surfaces wherein the planes of the playing surfaces intersect each other generally at the crown portion of the *racket* head or *paddle* blade. This relationship of the playing surfaces permits a greater arc of stroke by a player thereby permitting a struck ball to be propelled with a greater velocity. It also permits a player to strike a ball in a more forward position thereby making it easier for the player to "keep his eye on the ball," with a resulting better ball control.

This invention relates in general to new and useful improvements in sports equipment, and more particularly to novel game rackets and paddles with particular reference being made to tennis rackets and table tennis paddles.

In substantially all tennis rackets in use today, a single mesh of tensioned stringing lying in the median plane of the racket head frame constitutes the two playing surfaces of the racket. Also, substantially all table tennis paddles in use today are provided with a blade having two plane parallel faces which constitute the two playing surfaces of the paddle.

In the patents to Blache, U.S. Pat. No. 1,502,845, Daquan, U.S. Pat. No. 3,968,966, and Blackburn, U.S. Pat. No. 4,049,269, it is proposed to provide a racket with two discrete playing surfaces, each consisting of a mesh of tensioned stringing disposed in a plane generally parallel to the plane of the other, and separated from the other by the thickness of the racket head frame. It is an object of the present invention to provide an improved racket of this "double-strung" type but differing from that proposed by Blanche, Daquan, and Blackburn in that the planes of the two discrete playing surfaces are not parallel but instead, intersect each other and the longitudinal axis of the racket in the vicinity of the crown of the racket head. It is a further object of this invention to provide an improved paddle which differs from the conventional parallel-faced paddle in that the planes of the two playing surfaces (i.e., the paddle blade faces) are not parallel but instead, intersect each other and the longitudinal axis of the paddle in the vicinity of either the crown or the shoulder of the paddle blade. This nonparallel relationship of the playing surfaces permits a greater arc of stroke by a player thereby enabling a player to propel a struck ball with a greater velocity, or alternately, enabling a player to propel a struck ball satisfactorily with less physical effort. It also permits a player to strike a ball in a more forward position thereby making it easier for the player to "keep his eye on the ball" with a resulting better ball control. Furthermore, the angular relationship between the playing surfaces and the longitudinal axis of the racket or paddle will induce spin to a struck ball, the spin axis being generally normal to the plane of the stroke, and thereby causing the ball to "hook" or "slice".

United States Patent	4,293,129
Planks	October 6, 1981

Game rackets and paddles with non-parallel playing surfaces

Abstract

This relates to game rackets, such as *tennis* rackets, and to paddles, such as *table tennis* paddles, with two discrete playing surfaces wherein the planes of the playing surfaces intersect each other generally at the crown portion of the *racket* head or *paddle* blade. This relationship of the playing surfaces permits a greater arc of stroke by a player thereby permitting a struck ball to be propelled with a greater velocity. It also permits a player to strike a ball in a more forward position thereby making it easier for the player to "keep his eye on the ball," with a resulting better ball control.

Inventors:	Planks; Leo N. (Vienna, VA)
Family ID:	21913742
Appl. no:	06/040,924
Filed:	May 21, 1979

United States Patent 4,336,942

Warehime June 29, 1982

3-Way mini-tennis game and apparatus

Abstract

A 3-way *mini-tennis* game and apparatus therefor, wherein game is played by three players or teams on a hard, smooth, and level surface using a small resilient play ball, three or six paddles or rackets, and a special 3-way ball passage defining structure placed in centre of playing area on playing surface. The special structure in a preferred embodiment of the invention comprises three circular type hoops fastened together with stiff spring clips to form an upright regular pyramidal tripod, with sides of structure making a 60. degree. angle with playing surface, as measured within structure. Other optional shapes and materials can be used to form sides of structure, or a unitary, self-supporting 3-way ball passage defining pyramidal tripod structure can be used. No barrier nets are required. In general, game apparatus is safe, inexpensive, portable, and easy to set-up. Little or no boundary lines or court markings are needed. Singles or doubles can play game. In general, game play requires that each pass of ball, when hit by *paddle or racket,* go cleanly through two open areas of sides of sides of 3-way structure. Fault assessments and point score awards are somewhat like those for *paddle or table tennis*. The game can be played indoors or outdoors.

Inventors:	Warehime; Norwood R. (Baltimore, MD).
Family ID:	22698553
Appl. no:	06/189,733
Filed:	September 22, 1980

The invention as claimed is intended to provide a new approach to 3-way tennis type game play and apparatus therefor. It solves the problem of how to design a simple 3-way ball passage defining structure which represents the main central game apparatus (barrier) in game play. No nets are used.

The advantages offered by the invention are mainly that the elaborate net barrier structure is eliminated and that three identical frame elements, such as circular hoops, inverted "U" shapes, rectangle shapes, or the like, can be easily fastened together with spring clips to form an effective, inexpensive, and self-supporting 3-way ball passage defining tripod structure. An optional unitary tripod structure may also be used in place of the assembled tripod structure. In general, the novel game apparatus is safe, inexpensive, portable, and easy to set-up. Game court can be on any hard, smooth, and level surface, and requires little or no court layout markings. Game play by single or doubles is safe, dynamic, and suitable for a broad range of player types. Games can be played indoors or outdoors.

United States Patent	5,288,085
Young	February 22, 1994

Table racket ball game device

Abstract

This new invention is a *table Racket* Ball Game Device. This new invention is an enclosed *table* game device played with a *table tennis* type ball, off the sides, bottom and top, in three dimensions. This new *table* game device is a game of skill not chance. This is a serious game almost anyone can learn to play. It teaches eye-hand coordination and concentration. The ball is hand loaded, on to the *paddle,* aimed and carefully shot, at the other player's goal, with just a flick of the fingers. Players score points when the ball falls out of play behind the defender's defence and down the ball return ramp. The paddles are designed and mounted in such a way that the entire surface maybe used in defence. This is a game of skill not chance because the players can defend every shot if they can slide the *paddle* quick enough to block the oncoming shot.

Inventors:	Young; Robert G. (Lubbock, TX).
Assignee:	Young; Robert G. (Lubbock, TX).
Family ID:	21918745
Appl. no:	08/041,862
Filed:	April 2, 1993

United States Patent	7,204,770
Mathieu	April 17, 2007

Table tennis racket

Abstract

The present invention relates to a *table tennis racket* comprising a core of synthetic material and a covering fitted onto said core by overholding or by co-molding. According to the invention, the core is in the form of a single piece presenting a tubular handle and a *paddle of* thickness that is smaller than the smallest outside diametral dimension of the tubular handle, and it is provided with a plurality of recesses, between them defining a plurality of partitions extending perpendicularly to the plane of the *paddle.*

Inventors:	Mathieu; Stephane (Breteuil, FR)
Assignee:	Establishments Cornelia (Bonduel les Beaux, FR)
Family ID:	34955410
Appl. no:	11/333,237
Filed:	January 18, 2006

United States Patent Application 20160206954

Kind Code A1

Miller; Kenneth C. July 21, 2016

ROBOTIC GAME WITH PERIMETER BOUNDARIES

Abstract

A method and system are provided for identifying when and which robot targets fall into which pockets of a billiard *table* or pass through gates in the walls of a perimeter boundary, as well as well as boundary walls and gates that may be used to form various shaped playing areas. The identification of a robot target in a pocket may be used for scoring traditional games of pocket billiards, or for scoring robotic based games played on the surface of the billiard *table*. The identification sensor in certain embodiments may be mounted on the user configurable gates. Numeric values may be printed on the face of the robot targets to determine a winner of a game based on a total score of values of those robot targets knocked in by a player by themselves or by controlling a robot that pushes the robot targets.

Inventors: Miller; Kenneth C.; *(Aptos, CA)*

Applicant: Name City State Country Type

MILLER; Kenneth C. Aptos CA US

Family ID: 52587291

Appl. No.: 14/914772.

Filed: August 27, 2014

PCT

August 27, 2014

Filed:

PCT NO: PCT/US14/52908 371 Date: February 26, 2016

United States Patent Application	20180053114
Kind Code	A1
Ajoite; Akali	February 22, 2018

ARTIFICIAL INTELLIGENCE FOR CONTEXT CLASSIFIER

Abstract

An artificial intelligence system comprises a computer network server connected to receive and analyse millions of simultaneous texts and/or voice messages written by humans to be read and understood by humans. Key, or otherwise important words in sentences are recognised and arrayed. Each such word is contributed to a qualia generator that spawns the word into its possible contexts, themes, or other reasonable ambiguities that can exist at the level of sentences paragraphs, and missives. A thesaurus-like *table* is employed to expand each word into a spread of discrete definitions. Several such spreads are used as templates on the others to find petals that exhibit a convergence of meaning. Once the context of a whole missive has been predicted, each paragraph is deconstructed into sub-contexts that are appropriate within the overall theme. Particular contexts identified are then useful to trigger an actionable output.

Inventors: Ajoite; Akali; *(Mill Valley, CA)*

Applicant: NameCityState Country Type

BRIGHTERION, INC. San Francisco CA US

Assignee: BRIGHTERION, INC. San Francisco CA

Family ID:	61190756
Appl. no:	14/613383
Filed:	February 4, 2015

United States Patent Application	20190213498
Kind Code	A1
Ajoite; Akali	July 11, 2019

ARTIFICIAL INTELLIGENCE FOR CONTEXT CLASSIFIER

Abstract

An artificial intelligence system comprises a computer network server connected to receive and analyse millions of simultaneous texts and/or voice messages written by humans to be read and understood by humans. Key, or otherwise important words in sentences are recognised and arrayed. Each such word is contributed to a qualia generator that spawns the word into its possible contexts, themes, or other reasonable ambiguities that can exist at the level of sentences, paragraphs, and missives. A thesaurus-like *table* is employed to expand each word into a spread of discrete definitions. Several such spreads are used as templates on the others to find petals that exhibit a convergence of meaning. Once the context of a whole missive has been predicted, each paragraph is deconstructed into sub-contexts that are appropriate within the overall theme. Particular contexts identified are then useful to trigger an actionable output.

Inventors: Ajoite; Akali; *(Mill Valley, CA)*

Applicant: Name City State Country Type

Brighter ion, Inc. Purchase NY US

Assignee: Brighter ion, Inc. Purchase NY

Family ID: 61190756

Appl. No.: 16/299998.

Filed: March 12, 2019

United States Patent Application 20170348573

Kind Code A1

ZHOU; JIEBAO December 7, 2017

Table tennis racket rubber and wood staggered stack handle design

Abstract

The present invention relates to changing three aspects of *table tennis rackets. Table tennis* racket rubber components of this invention are crossly inserted between racket handle small side board and bottom board. The racket handle is increased from the current three layers to five-layer structure; The material of the racket handle is changed, and the rubber elastic layer is added. The handle of the racket is composed of wood and rubber, forming a flexible racket handle; It adopts rubber (or soft material) safety device to cover the exposed part of the racket bottom board, reduces the friction between the palm and the bottom plate, and enhances comfort; The invention improves the safety, comfort and controllability of the racket handle and accelerates the speed, improves the accuracy of *table tennis*.

Inventors: ZHOU; JIEBAO; *(SOUTH EL MONTE, CA)*

Applicant: Name City State Country Type

ZHOU; JIEBAO SOUTH EL MONTE CA US

Family ID: 60482888

Appl. No.: 15/669991.

Filed: August 7, 2017

United States Patent Application	20150258400
Kind Code	A1
Yamamoto; Aaron	September 17, 2015

Racket sport system and method of play for an aerial projectile game

Abstract

A racket sport system and method of play for an aerial projectile game are disclosed. The system generally comprises of two truncated-cone or pyramid shaped containers, at least two *rackets,* and a game ball. The containers are targeted objects with an opening located on the top and on the front sidewall for game ball entry. Players of two-player teams are placed a distance apart as with the distance apart of the two containers. Teams take turns serving and receiving the game ball with their *rackets* to contact or enter the opposing containers for assisted or unassisted points. To win, a team must be the only team to score exactly twenty-one points or score the most points in the overtime round(s) when a tie occurs. Teams can also win immediately during any round when a player serves the game ball into an opposing container without assistance from their teammate.

Inventors: Yamamoto; Aaron; *(Mesa, AZ)*

Applicant: Name City State Country Type

Yamamoto; Aaron Mesa AZ US

Family ID: 54067868

Appl. No.: 14/633089.

Filed: February 26, 2015

United States Patent Application		20120214625
Kind Code	A1	
Brandt; Richard A.	August 23, 2012	

TENNIS RACKET AND METHOD

Abstract

A *tennis* racket is provided that has a rectangular ball striking face with rounded corners. The racket is formed of a carbon fibre sandwich with light material mounted therebetween. A strip of carbon fibre is attached to the outer edge spanning the two-carbon fibre top and bottom sheets. The rectangular frame is formed with outwardly bowed sides so as to be pulled into a straight sided configuration by the string tension. Locking grommets of various constructions are provided to hold the strings at predetermined tensions. Testing equipment and methods are provided for testing *tennis rackets.*

Inventors: Brandt; Richard A.; *(New York, NY)*

Family ID: 46581119

Appl. No.: 13/352853.

Filed: January 18, 2012

United States Patent Application 20020098925

Kind Code A1

Brandt, Richard A. July 25, 2002

Sports racket having a uniform string structure

Abstract

A sports racket, for *tennis* and the like sports, has an elongated handle attached to a head having a racket face, which is spanned by a uniform string structure. The head has four sides, forming a non-elliptical shape, in which the opposite sides are substantially parallel. Longitudinal strings, all of which are substantially identical in length and run essentially parallel to each other, and transversal strings, all of which are substantially identical in length and run essentially parallel to each other and perpendicular to the longitudinal axis, span the racket face. The racket has a larger racket face than conventional *rackets,* while maintaining the length and weight measurements of conventional *rackets,* resulting in a very large sweet spot and more good hits. The racket has maximally long strings at all points on the racket face and strings substantially identical in length at all points on the racket face resulting in a uniformity of response for off- centre hits, an increase in the ball rebound speed, a decrease in angular deflection for off-center hits, and the ability to set the tension of the strings such that they vibrate with the same frequency. The racket has a greater moment of inertia than conventional *rackets,* resulting in reduced racket rotation and a reduction of injuries to players, such as ""*tennis* elbow"".

United States Patent	5,910,528
Fali off, et al.	June 8, 1999

Inventors:	Fali off; Wahida (Ashland, OR), Sander; Harald (Ashland, OR)

Family ID:	21765587
Appl. no:	08/826,163
Filed:	March 27, 1997

Table tennis rubber solvent and adhesive systems

Abstract

Solvent systems, and adhesive systems comprising such solvent systems, primarily for use in forming speed-glued *table tennis rackets* are described. One embodiment of the solvent systems comprises a cycloalkane having from about 3 to about 10 carbon atoms, an ester having from about 2 to about 10 carbon atoms, and a terpene having a molecular weight of from about 110 to about 160. The cycloalkane is generally selected from the group consisting of cyclopentane, cyclohexane, and cycloheptane, with cyclohexane being a currently preferred cycloalkane. A currently preferred ester is ethyl acetate, and a currently preferred terpene is limonene. The solvent system may also comprise an ether, and preferred ethers are Di propylene glycol dimethyl ether and propylene glycol methyl ether acetate. The solvent system can also include an aliphatic alkane having about 10 carbon atoms or fewer, the aliphatic alkane being used in sufficient quantity to adjust the vapour pressure of the solvent system to be from about 50 mm/Hg to about 100 mm/Hg at 25. degree. degree. C. Examples of suitable aliphatic alkanes are heptane, 2,2,4-trimethylpentane and mixtures thereof. The present invention also provides adhesive systems for use in forming speed-glued *table tennis rackets*. The adhesive system comprises the solvent systems described above, and from about 2 to about 10 weight percent of a solid or solids suitable for speed gluing *table tennis* rubbers to *table tennis* blades. A method for speed gluing a *table tennis* racket is also described. The method generally comprises forming a solvent or adhesive system as discussed above. The solvent or adhesive system is then applied to a *table tennis* rubber, a *table tennis* blade, or both. The rubber is then attached to the blade to form a speed-glued *table tennis* racket. If just a solvent system is used, then the rubber and/or the blade must provide sufficient solid or solids to adhere the rubber to the blade.

Printed by Books on Demand GmbH, Norderstedt / Germany